IT TAKES ALL KINDS

WRITTEN BY SALLY MARKHAM-DAVID
ILLUSTRATED BY TREVOR RUTH

Contents

Introduction

An animal's shape reflects where and how it lives and the way it moves. On land, some animals have bodies shaped for speed, so they can move fast to catch their food or to escape from their enemies. Some animals can move easily through the tree canopy, swinging gracefully from tree to tree. Other animals have bodies that are best suited to swimming in water. Birds have wings that are shaped for speed, and for soaring and gliding. They have light bones so that their wings can support them while flying.

The size and shape and appearance of an animal has a lot to do with where and how it lives. The coat will be of a kind to suit the climate. The feet will be adapted to the kind of ground it walks on. The mouth will be shaped to fit the kind of food available.

Sometimes, animals that live the same kind of life, like the flying squirrel and the sugar glider, may develop the same body shape even though they are not related.

Others, even though they belong to the same family, like the bear and the sea lion, have evolved quite different shapes because one lives on land and the other spends a lot of time in the sea.

Runners

The cheetah is the fastest animal in the world. When it hunts, it can sprint in short bursts at speeds of up to 60 miles per hour. Its supple backbone bends like a spring, giving an extra thrust to each stride.

The antelopes and gazelles that are often the cheetah's prey need to be speedy, too. They run on the very tips of their toenails, which are their hoofs. They can change direction much quicker than the cheetah and can often dart away out of danger.

Swingers

Animals that swing through the treetops usually have hands and feet that can grasp branches. Some animals, such as spider monkeys and some possums, can use their tails as an extra arm.

Leapers

Animals that can leap, such as kangaroos, grasshoppers, and frogs, all have strong back legs with large thigh muscles. A kangaroo has long feet and its body is balanced with a long tail. It can leap up to 25 feet in a single bound.

Many other animals that leap, such as hopping mice, also have long feet and a long tail for balance.

Sliders

Some animals, such as snakes, have no limbs at all. They slither along the ground as muscles ripple through their bodies.

Snails also move by rippling the muscles along their flat underside, or foot.

Swimmers

The best shape for swimming easily and fast through the water is the streamlined shape of a fish. Mammals that live in the water, such as seals, dolphins, and whales also have this streamlined shape.

Fliers

There are many animals, apart from birds, that have developed body features designed for flight. Flying fish have winglike fins that allow them to make long flying leaps above the water.

Some squirrels and possums have flaps of skin that stretch from their wrists to their ankles. When they stretch out their legs these flaps enable them to glide and parachute from one tree to another.

It All Depends on Where You Live

The camel has adapted to the harsh climate and the sandy soil of the desert. The camel's body heats up slowly and cools down quickly, and its long legs keep it well above the hot sand during the day. The camel's thick, wooly coat insulates it against the heat of the day and keeps it warm on cold desert nights.

A camel can take in a large amount of water and then go without a drink for days. As a camel sweats very little, it does not lose much water. Also, a crease from the nostrils down to a split in the upper lip forms a channel allowing mucus from the nose to be 'drunk.' No water is wasted!

The hump on the camel's back is a store of fat which the camel can use as an extra food supply when necessary.

The camel's feet have large, thick pads that help it to walk on sand. When the wind blows the sand about, the camel can close its nostrils to keep the sand out of its nose, and its long eyelashes protect its eyes.

Gibbons, the smallest of the apes, have a body shaped to suit their life in the forest. Their arms are much longer than their legs and can move freely in all directions. Gibbons swing through the trees, arm over arm, with their long fingers hooked over the branches. Their bodies are light and agile, and they can leap from tree to tree.

On the ground, gibbons often walk upright, holding out their arms for balance. Their diet is mostly the fruits and nuts of the forest, but they also eat leaves, insects and small birds.

Same Family, Different Address

Some animals who belong to the same family seem at first glance to look alike. They have the same body shape, but they have developed special features to suit the places where they live.

In the turtle family, those that spend most of their time on land have legs with webbed feet and claws. They are called tortoises.

The limbs of turtles that spend most of their time in the water have developed into flippers.

Sea turtles also have flatter shells which are more streamlined for swimming. Some turtles can take in oxygen through patches of skin on the throat, so they can stay underwater longer.

The kangaroo and the tree kangaroo appear to be alike, but the tree kangaroo has evolved special features to suit its life in the trees. Whereas the kangaroo has large feet for leaping, tree kangaroos are smaller and have shorter hind feet. The pads on their feet are broader and have cushioning soles covered with rough skin, like non-skid tire treads.

The hands of the tree kangaroo are larger in proportion to their body size than any other kangaroo, and they have strong curved nails which are good for gripping.

The tail of the tree kangaroo has developed differently, too — it is longer and more slender. It has a furry tip which acts as a rudder during long aerial jumps, and as a prop when climbing.

Stranger Than Fiction

Some animal bodies look as though they were put together at a lost-and-found—a mix and match of whatever happened to be on the shelf.

The aardvark, an anteater of South Africa, certainly looks as though it was made out of spare parts. Its name means earth pig, and it does have a snout like a pig's. But the aardvark also has the ears of a donkey, the tail of a kangaroo, and the body of a bear.

Yet the aardvark is perfectly formed for its way of life. This animal lives in a burrow. An aardvark's front feet make such good digging tools that it can dig faster than a team of six people with spades. The aardvark's main food is termites, and its feet can rip through the walls of termite nests that would be hard to break with a pickax. Like the giant anteater, the aardvark has a long, sticky tongue for capturing its prey, and bristly nostrils which can be closed to keep the termites out of its nose.

The platypus, found only in Australia, is the only animal in the world that has hair like a mammal, and webbed feet and a bill like a duck. The platypus lays eggs, like a duck, yet feeds its young with milk, like a mammal. It swims under water to find all its food, but still needs to come to the surface to breathe.

Platypuses sleep and nest in burrows which they dig in the banks of the rivers or lakes where they live. The entrances to the burrow are always above the water so the nest remains dry.

The platypus is well suited to its life in and out of the water. When the animal is on land, the webbing on its front feet can be turned back so that the claws are free for digging. The eyes and ear openings close under water, but receptors located on the top of the bill are so sensitive that the platypus can find its food by touch.

Disguises

Is it a twig or an insect, a treehopper or a thorn, a leaf or a bug? In the endless struggle to survive, animals have taken on many disguises. Many creatures that have no other way to defend themselves use a disguise, or camouflage, to confuse their enemies.

The stick insect is invisible as it lies amongst the twigs.

Treehoppers, sucking the juice along the stem of a plant, look so much like thorns that they deceive the birds.

The Indian leaf bug looks just like a leaf, right down to the veins. It even waves in the wind the way a leaf does.

Index